There are queen cells in my hive -

what should I do?

by Wally Shaw

Northern Bee Books

There are queen cells in my hive - what should I do?
© Wally Shaw

ISBN 978-1-912271-17-7

Published by Northern Bee Books, 2018
Scout Bottom Farm
Mytholmroyd
Hebden Bridge HX7 5JS (UK)

Design and artwork by D&P Design and Print
Printed by Lightning Source UK

There are queen cells in my hive -

what should I do?

by Wally Shaw

There are queen cells in my hive - what should I do?

There are queen cells in my hive - what should I do?

Introduction

You have opened a hive and found queen cells and the knee-jerk reaction (more of a hand-jerk really) of many beekeepers is to destroy them under the misguided notion that this will prevent swarming. So the first rule is 'get a grip' and **DON'T PANIC!** Destroying queen cells to prevent swarming never has been and never will be a successful method of swarm control. If you destroy one lot of queen cells the colony will immediately make some more and will probably swarm earlier than normal in their development - often before the first queen cells are sealed. If you destroy queen cells twice you run the risk of the colony swarming and leaving behind no provision for a new queen. Any delay in swarming that you induce by destroying cells will often result in the prime swarm being larger than it would have been if you had not interfered. Once a colony of bees is triggered to swarm it is very rare for them to go off the idea. At this point the beekeeper must take control of the situation by using an appropriate form of management. For an attentive beekeeper (who does regular inspections to check for signs of swarming) the queen cells will usually be sealed before the colony has issued a prime swarm and making an **artificial swarm** is what is required. If the colony has already swarmed then other management techniques can be used to prevent a cast swarm and the loss of further bees but the details depend on the current status of the colony. Approaching the problem logically and finding out exactly what stage of the swarming process has been reached will give the beekeeper the best chance of successfully intervening; **thereby not losing bees, saving as much of the potential honey crop as possible and not ending up with a queen-less colony.**

Just occasionally queen cells are torn down and the colony seems to go off the idea of swarming. It is said that this can happen naturally if an abundant flow of nectar occurs or that it can be artificially induced by removing several frames of brood and replacing them with foundation. The intervention of a nectar flow is outside the beekeeper's control and the `shock` removal of brood and the introduction of foundation may work occasionally but can certainly not be relied upon.

Hive Diagnosis

Before you contemplate any management of a colony that has developed queen cells you first have to understand what is going on – are the queen cells there for swarming and if so what stage in the process has the colony reached? All the information you need to make this diagnosis is `written` on the brood combs and to a lesser extent the bees. But before you make this diagnosis you have to know how to `read` the combs – understanding what you can see and what else you need to look for to reach a conclusion. In many ways this is like a forensic investigation. To be able to do this effectively you must have a basic knowledge of honey bee biology and behaviour.

Reasons for Queen Cells Being Present in a Hive

There are three different reasons why a colony produces queen cells:-

1) Reproduction - swarming
2) Replacement of the existing queen - supersedure
3) The colony is queen-less – emergency re-queening

These are three separate and distinct behavioural programmes but nine times out of ten (or probably nearer 19 times out of 20) when a beekeeper opens a hive and sees queen cells it is the swarming programme that they are following. We often refer to the 'types' of queen cell because the origin of the cell and it larva, its position in the hive and the number of cells that have been produced are characteristic of the programme that has been activated. These are important clues as to what is going on but not completely definitive (see **Ambiguous Situations** below).

Before deciding on any management in response to finding queen cells it is necessary to correctly identify which programme the colony is on. **Figures 1-3** show typical queen cells (external appearance and position) for the three programmes. Only the presence of swarm cells means that the colony is intent on swarming and those produced for other reasons (supersedure and emergency re-queening) will on NO account result in the issue of a swarm. Supersedure and emergency queen cells do not usually require any intervention from the beekeeper – **except to leave the bees well alone and let them get on with it.** So, how do you know which programme the colony is following?

1) Swarm Cells

These are developed from queen cups into which eggs have been laid (or transferred) and are entirely vertical. The cells are long and usually constructed on the edges of the frames; either along the bottom bars or in recesses on the sidebars (see Figure 1). The longest cells are based are based on a boss of wax which enhances their apparent size but the internal volume of queen cells (which is what really matters) is very uniform. Occasionally some cells will be on the face of a comb - particularly when the brood nest is in a single box. In terms of number, there are rarely less than 5-6 queen cells, more typically 10-20 and possibly up to 100 of them. As the name suggests, the colony is producing new queens so that it can swarm and unless the beekeeper intervenes this is exactly what will happen. Swarming will usually occur around the time that the first queen cells are sealed **(day 8)** but it can occur earlier, especially if the beekeeper has previously destroyed queen cells. Swarming can also occur early for no apparent reason – some colonies just do! It can also occur later if delayed by poor weather and, in extreme cases, the new queens may be ready to emerge from their cells but `warder` bees are keeping them imprisoned until their mother has departed with the prime swarm.

Figure 1: Swarm cells in typical position

2) Supersedure Cells

Like swarm cells, these are typically vertical and are usually located on the face of the comb (see **Figure 2**). But they can also have the same origin as emergency queen cells (see below) and be based on an egg laid in a worker cell (not a queen cup) – they look like an emergency cell but are there for a quite different purpose. In supersedure there are usually only 2-3 cells grouped together on the same comb. The intention here is to replace the existing queen who they have decided is no longer up to the job. She may be old, she may be damaged and probably a host of other things of which we are not aware. Unfortunately one of the queen's defects that the bees seem unable to detect is when she is running out of sperm and is destined to become a drone-layer.

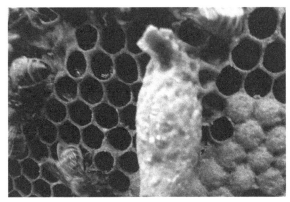

Figure 2: Supersedure cell

The books say that during supersedure the old queen is normally retained until her replacement has successfully mated and started to lay. In some cases, mother and daughter co-exist in the hive for some time but eventually the old queen will 'disappear'. This is called a 'perfect supersedure' but with the type of bees we have in Wales 'perfection' seems to be the exception rather than the rule. More usually the supersedure is 'imperfect' and there is a brood gap resulting from the old queen having been eliminated before her daughter comes into lay. If supersedure cells are found in a colony there is nothing the beekeeper needs to do except leave well alone and hope the outcome will be successful. Early spring and late autumn attempts at supersedure are often unsuccessful – lack of drones is the usual cause - and the situation needs careful watching to see that the colony does not become queen-less.

3) Emergency Queen Cells

These are queen cells produced in response to the sudden loss of the queen. Their identity is unambiguous because there will have been no queen in the colony since the cells were started and the younger stages of brood (and particularly eggs) will be missing. As the name implies, emergency queen cells are produced in response to the worst possible situation in which a colony can find itself (with no queen) and the only aim is to get a new queen as soon as possible – the last thing the colony wants to do is to swarm. The queen may have died suddenly of natural causes or the beekeeper may have killed her or spilt her onto the ground during hive manipulations. Emergency queen cells are also produced if the beekeeper deliberately removes the queen from a colony. Occasionally a colony may lose its queen more than 4-5 days after she last laid any eggs and in this circumstance there will be no brood young enough to make an emergency queen. Without intervention by the beekeeper this colony will be unable to re-queen itself.

There are usually numerous emergency queen cells (see **Figure 3**) but instead of being developed from an egg or larva in a queen cup, they are based on existing eggs or young larvae in normal horizontal worker cells. Nurse bees start to feed the selected occupant with royal jelly and the outer rim of the cell is extended downwards to make room for the increaseing size of a queen larva. There seem to be two types of emergency queen cell; one type has the surrounding comb extensively modified to produce a vertical cell (very similar to a swarm cell but inset on the comb face – see **Figure 4a**) and the other type is part horizontal and part vertical with a right-angle bend in the middle (see **Figure 4b**). Although a bent cell (externally) looks inferior, both types seem to contain queens that are indistinguishable in terms of size. The fully vertical cell is most common on new comb and the right-angled version on old comb – presumably because of the difficulty of nibbling away cell walls with several layers of tough pupal skins.

At first sight emergency queen cells look rather unimpressive when compared with swarm cells - and they can be easily overlooked unless the bees are shaken or brushed from the frame. There is a firmly rooted myth in beekeeping that queens developed in emergency queen cells are inferior to those from swarm cells. Despite, from the outside, looking smaller than swarm cells, emergency cells normally produce perfectly good queens. The idea that emergency queen cells produce an inferior queen (a `scrub` queen) is probably based on last-ditch attempts by beekeepers to re-queen colonies that have been queen-less for some time by giving then a frame with eggs or young larvae on it. Such colonies are populated by OAP's and simply do not have enough nurse bees of the right age to produce a fully developed queen.

Figure 3: Emergency queen cells

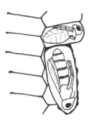

Figure 4a: Fully vertical emergency queen cell

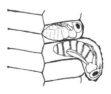

Figure 4b: Right angled emergency queen cell

Ambiguous Situations

In the vast majority of cases when a beekeeper opens a hive and finds a significant number of queen cells the intentions of the colony are completely obvious – it is **SWARMING** and there is no room for confusion. However, it should be emphasised that it is not the appearance of queen cells or their position on the frame that is most important, it is why they are there in the first place; in other words, the behavioural programme that the colony is following. The question sometimes arises, is it swarming or supersedure (?) and this is not always obvious. For example, swarm cells are not always on the edges of frames and supersedure cells are not always on the face of the frames and, just to confuse matters, the number of cells may be atypical – too few to be swarming or to many to be supersedure. Fortunately emergency re-queening is always obvious because there will be no eggs in the hive and the youngest brood will tell you exactly when the queen was lost.

So the most common problem is distinguishing between swarming and supersedure. In swarming there are usually multiple queen cells present but what is happening if you find just 3-4 queen cells on the bottom bar of a frame – which programme is this? If the cells are on more than one frame it is more likely to be swarming but if they are on the same frame it is probably supersdure – but you can't always be sure? Similarly, more than 3 queen cells on the face of a frame is usually regarded as too many cells to be a supersedure – but it may be! If the queen cells are newly started and will not be sealed for several days (and there is no immediate risk of a swarm) the decision what is going on can be deferred. The beekeeper can come back in 2-3 days time and if more queen cells have been started it is almost certain that swarming is their intention. If there are no additional queen cells then it is probably a supersedure. If the cells are sealed (or close to sealing) and swarming is potentially imminent and you can't be sure **it is best to play safe and a 'belt and braces' method of dealing with this situation is described in Appendix 1.**

There are some circumstances where swarm and emergency (type) queen cells can co-exist in a colony. For example, if a colony swarms early (before the swarm cells are sealed), the workers will usually respond to what they perceive as loss of the queen by making some additional emergency cells. This is a direct response to the sudden loss of queen pheromones in the colony and bees that are responsible seem to be unaware that they already have numerous swarm cells present. The same thing can happen if the beekeeper removes the queen when making an artificial swarm. In neither of these cases has the colony changed to an emergency re-queening programme and the emergency cells are usually of no practical significance; they are so much younger than the swarm cells and are unlikely to survive to maturity. However, there is one situation where emergency queen cells can matter and that is when a colony has already swarmed and the beekeeper has been forced into culling all but one (1) of the remaining queen cells to prevent the issue of a cast swarm. If the culling is done immediately after the prime swarm has departed there will still be eggs and young larvae present from which the bees can make emergency queen cells. These cells will be started after the beekeeper has closed-up the hive satisfied that everything is under control. Being in swarming mode, the colony may proceed to cast swarm with the queen that emerges from the queen cell the beekeeper left intact, treating the emergency cells as backup to provide their new queen. Again it is not the type of cell that matters but the behavioural programme the colony is on that determines the outcome.

Some Other Basic Facts you Need to Know

The developmental stages of the three types of brood (queen, worker and drone) and timing (in days) are shown diagrammatically in **Figure 10** (see page 28). In order to understand what you are looking at in the hive you need to familiarise yourself with the key developmental stages, their appearance and their timings.

Queen Cell Development

The earliest you can identify a viable queen cell is when it is already 3 days old - an egg in a queen cup does not necessarily mean it will become a queen cell. The critical decision for the colony is made when the egg hatches out (**Day 3**) and the nurse bees start to feed the larva with royal jelly. A queen cup with a pool of royal jelly and a tiny larva in it will almost inevitably be taken full term to become a sealed queen cell. Sealing takes place on **Day 8**, ie. the larval feeding period is just 5 days. Once queen cells are sealed it is difficult to know how old they are without breaking a few open to take a look at the contents. There are usually cells covering a range of ages present, so if you want to know the timing you need to look at cells in several in different parts of the hive to ensure you have covered all eventualities.

Emergence of queen cells occurs on **Day 16**, ie. 8 days after sealing. A newly emerged queen cell usually has a hinged lid attached (sometimes it may fall off) but it is also quite common for the bees to close the lid and reseal it – look for a line round the tip of the cell (a sort of `tear here` line). You may be surprised to find an occupant in such cells but usually it is a worker bee that has gone in to do a bit of cleaning work and has been sealed in by some tidy-minded sister. If the bee is head-down in the cell it will be a worker but, if it is head-up, it will be a queen and she will just be waiting for you to open the cell to walk out. Again, **don't panic and kill her!** Simply let her walk off into the colony because this is an extremely easy (even advantageous) situation to resolve (see **Step 7** below).

Worker Brood Development – If when you open the hive you find sealed (or point of seal) queen cells the colony may already have swarmed. You may already have some clue that this has happened from the number of bees in the hive being less than you expected. Confirmation that the colony has not yet swarmed is to find newly laid eggs (standing on end in the bottom of the cell) or, better still, see the queen herself. If there are only eggs lying down in the cells then this is indeterminate and you really need to see her (the queen) to be absolutely sure. If there are no eggs then the colony has almost certainly swarmed and the age of the youngest larvae will tell you when this happy event (happy for the bees but less so for you) occurred.

Eggs hatch on **Day 3** and worker brood is sealed on Day 9 (emerges on day 21), so there are 6 days of feeding and the bigger the larva the older it is (see Figure 10). The reason you need to know when the colony swarmed is so that you can assess how imminent is a cast (or secondary) swarm. It is also useful to know when the swarming actually took place so that you understand how you managed to miss seeing the warning signs – this an important learning opportunity! If there are no unsealed larvae in the colony – only sealed brood - then it is at least **9 days** since the colony swarmed and you are in serious trouble because a cast swarm is imminent or has already occurred. For how best to recover from this situation, see Steps 4 and 5 below.

Drone Brood Development – Drone brood is sealed on **Day 10** (emerges on day **24** or even as late as day **28**). There is not much to be said about drone brood except do not rely too much on the information that can be gleaned from its stage of development. In a swarming situation, drone brood is the first to be neglected by the worker bees; it may be poorly fed, remain unsealed for long periods and even die. Unsealed worker brood is also subject to a higher mortality rate in a queen-less, post-swarming colony.

Diagnostic Tree and Remedial Management

The main period for swarming is May-July with a peak in late May and most of June. Outside this period swarming is less likely but can still happen. It is possible for a colony to go from no obvious signs to actually issuing a swarm in 5 days - or less if they go before any queen cells are capped. In order to catch swarming in the early stages, regular hive inspections are required but how often should these be done? It is difficult to lay down hard and fast rules because it depends on your knowledge of your bees, the area in which you keep them and, perhaps more importantly, what time you have available to 'mess around' with your bees. Swarming is also highly dependent on weather; good weather with a nectar flow tends to suppress the swarming urge and poor weather with little flying time tends to promote it. As a rough guide, during periods when there is increased risk of swarming, inspections should be done every 5 days. When the risk is considered to be lower, 7-10 day inspections will probably suffice.

A well-known method of extending the need to check for queen cells to 14 days is to clip the queen's wings thus preventing her from being able to fly far. When a colony with a clipped queen tries to swarm the queen is not air-worthy, falls to the ground, the bees cluster around her for a while but eventually return to their hive in disgust and await the emergence of a virgin queen with whom they will swarm. Knowing that the queen has clipped wings should enable the beekeeper to immediately identify what has happened and apply the appropriate management to prevent cast swarming (see **Step 5 or Step 7** below - your choice as to method). There are pros and cons to the practice of clipping queens the discussion of which is beyond the scope of this booklet. Failure of the prime swarm can also occur naturally when for some reason an intact queen is unable (or unwilling) to fly. To identify this situation and prevent the colony re-swarming with a virgin queen remedial management is required (see **Appendix 2**).

One of the advantages of keeping bees on a two-box system (brood and a half or double brood) is that queen cells will usually be started on the bottom bars of the upper box. If no other management is required, a swarm check can be very quickly accomplished by simply lifting one side of the upper box and looking for signs of queen cells on the bottom bars (use smoke to move the bees and get a proper look). This type of inspection is not 100% reliable but good enough in most purposes – and certainly a lot better than not looking at all!. Single brood box systems will require the removal of at least some frames to check for queen cells.

Regular inspections (swarm checks) will ensure that you never need to use the later steps in this diagnostic tree that follows.

Steps 1-3 cover pre-swarming development, with Step 3 putting the beekeeper on amber alert. Steps **4-9** deal with increasingly more advanced stages in the swarming process and **Steps 10-12** deal with problems that may arise after the swarming process is over and the colony has not returned to normal with a laying queen – in other words, a rescue programme. Each step consists of two parts; **Investigation** - instructions to help you to identify what stage the colony is in and **Remedial Action** - what to do about it (management required).

STEP 1 – There is drone brood in my hive.

Investigation – None necessary. This step is included because it is a widely held myth that the presence of drone brood means that the colony is preparing to swarm. The presence of drone brood is merely an indication that the colony has reached a certain stage in its spring build-up when it can `afford` to produce and support drones. All healthy, successful colonies produce drones as part of their normal development, usually starting in mid-March and continuing until sometime in August. Many of these colonies with early drone brood will make no attempt to swarm during the season

Remedial Action – No action is required. Just rejoice that the colony is developing normally but be aware that the presence of drone brood means that the Varroa mite population will start to grow more quickly and now is the time to check that the colony does not have a mite problem that will only escalate as the season progresses.

STEP 2 – There are queen cups in my hive.

Investigation – Check to see there are no cell contents; no eggs and particularly no young larvae in a pool of royal jelly.

Remedial Action - If there are no contents no action is required. Like drones, the building of queen cups (practice cups or fun cups), mostly on the bottom bars of frames, is a natural stage in the build-up of the colony and does not mean that swarming is imminent. Cup building happens because the queen is no longer regularly walking on the edge of the frames and leaving her footprint pheromone there - presumably because she is too busy with other matters and the hive is also becoming more congested.

STEP 3 – There are queen cups with standing-up eggs in them in my hive (see **Figure 5**).

Investigation – Check that no cells have gone a stage further and contain a larva in a pool of royal jelly.

Remedial Action – If there are only standing-up eggs no action is required except to go onto **amber alert** - this may be the start of something more serious. However, many colonies will have eggs in queen cups several times during the season and still make no attempt to swarm.

Figure 5: Queen cup with standing up egg

STEP 4 – There are queen cups with contents (larvae and royal jelly) in my hive and some of the cells are starting to be extended
(see **Figure 6**).

Investigation – This is a sure sign that the colony is almost certainly going to swarm - so it is **red alert** time. Now you need to find what stage of development the swarm cells are at so that you can estimate the time to their being sealed and therefore the likelihood of swarming in the near future. Just occasionally a colony will swarm prematurely – before there are any sealed swarm cells – so it is as well to check this has not already happened. Is the colony smaller than you expected, are there newly laid eggs or, better still, can you see the queen? If you decide is has already swarmed you need to go to the **Step 5**.

Remedial Action – If all the cells are in an early stage of development then you probably have time on your side (1-3 days?).But **do not procrastinate;** remember that some colonies swarm prematurely. So, in reality, you have to prepare to do an artificial swarm on the colony as soon as possible. There are many methods of artificial swarming to be found in beekeeping books. However, WBKA has published a booklet entitled 'An Apiary Guide to Swarm Control' and it is recommended that you use the method described there and accompanied by colour coded diagrams (**Snelgrove II modified**). As the name suggests the booklet is intended to be taken into the apiary as an 'on-the-job' source of information (a crib sheet). If you have no previous experience of doing an artificial swarm you might wish take advantage of some help from a mentor or a more experienced beekeeper if you can talk one into assisting you. Snelgrove II (modified) has the advantage that for the initial manipulation in making an artificial swarm **the queen does not have to be found.**

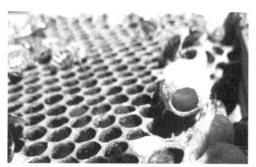
Figure 6: Queen cup with contents

STEP 5 – There are sealed queen cells in my colony
(see **Figure 7**).

Investigation – Now you must seriously consider whether or not the colony has already swarmed. The first clue is the number of bees in the hive. Are there fewer bees than when you last looked at them? Are the supers full of bees? Next look at the brood – are there newly laid eggs? Can you find the queen? If there are no eggs, what is the youngest brood you can see? If you can find standing-up eggs or have seen the queen you are in luck and the colony has not yet swarmed – **but it may do so at any moment if the weather is favourable and it is not yet 16.00 hours.** If you decide it has not swarmed, go back to **Step 4** and make an artificial swarm. If you decide it has swarmed then there is only one option open to you.

Remedial Action – If you judge that the colony has issued the prime swarm, all you can do now is to prevent a cast swarm and the further loss of bees – and with it any chance of a decent honey crop from this hive. The normally accepted method of doing this is to thin out queen cells until the bees have no option but to settle down with the only emerging queen. But what rule do you follow? There is a range of opinion on this matter. The most common advice is that you select an unsealed queen cell in which you can see a healthy larva and destroy **ALL** the rest (sealed and unsealed). If there are no unsealed cells you will have to settle for a sealed one – the best you can find and preferably one in a well-protected position so there is no chance of accidental damage during manipulation. Personally, I do not see the point in the unsealed cell option; bees will not seal queen cells when the occupant is already dead and all you are doing is foregoing the advantage of a sealed cell that will hatch earlier. Some queens die in the cell after it has been sealed and you can not avoid that possibility - but you must be aware it can happen and be prepared to take remedial action. Some writers recommend taking out insurance and leaving 2 cells of the same age – but how do you know cells are of the same age(?) - and using this method there is still potential for a cast swarm. **The golden rule of this recovery operation is that you MUST destroy ALL queen cells apart from the one (or two) that you select.** To do this properly it is usually necessary to gently shake or brush bees from the combs so that you can see what is there. Be particularly careful not to miss any queen cells in awkward corners at the bottom of the combs and tucked in to the sidebars - failure is not an option! Some books say you must not shake frames or you will dislodge larvae or pupae in the queen cells. However, as long as you shake gently there should not be a problem. If the colony has only recently swarmed (less than 4 days), there will be eggs and/or young larvae present from which emergency queen cells can be made. You will need to return in 3-4 days time and check that none have been started as this could undo your previous good work, enabling a cast swarm to occur.

A better option for the more experienced beekeeper - one that does not require thinning queen cells - can be found in **Step 7**. We have used this method exclusively since the first edition of this booklet was written and so far it has been 100% successful. Above all it avoids the beekeeper choosing who is to be the new queen and allows the bees to do it for themselves.

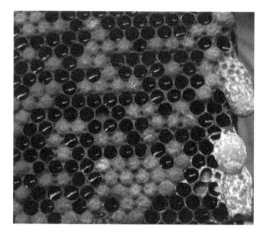

Figure 7: Sealed and unsealed cells

STEP 6 – My hive has definitely swarmed and is left with a reduced number of bees, brood and numerous queen cells.

Investigation – This colony will almost certainly produce a cast swarm if you do not do something to prevent this happening. How urgent the matter is depends on how long ago swarming occurred and how mature the queen cells were at this time. If you saw the hive swarm or caught a swarm which you know came from this hive then you are already in possession of the information you require. If you do not have this information you can find out by looking at the brood. Search for the youngest brood (and there may be very little of it so look diligently) and count back to day zero and that will tell you roughly how long it is since the swarm took place. How mature are the queen cells? You will find this out as you start to thin them (see Remedial Action, **Step 5**). If you find that queen cells are starting to emerge as you look through the hive, go straight to **Step 7** because this is a 'golden opportunity'. Or if the cells are not mature you can choose not to thin cells. Instead, open a few queen cells, estimate a date when they will be mature and again go straight to **Step 7**. There seems to be a 2-3 day window (a safety margin) after the first virgin has emerged and before the first cast swarm issues. Because the remaining (non-emerged) queens are likely to be more mature at the time of their release, this is probably not the case for subsequent cast swarms.

Remedial Action – Same as for **Step 5**.

STEP 7 – My hive has swarmed and there are emerged and sealed queen cells present.

Investigation – This sounds like a rather tricky situation but this is not the case. If the hive has already cast swarmed it has happened and there is nothing you can do about that – it is also difficult to be sure unless you have seen or caught the swarm. The best indication is a marked reduction in the number of bees in the colony. If it has not cast swarmed then you are in luck and it is usually possible to prevent this happening.

Remedial Action – Examine the remaining sealed queen cells which are probably on the point of emerging anyway. You may find that queens start to emerge from their cells as you look through the hive. This happens because your blundering around has distracted the `warder` bees that were keeping queens penned in until the colony wanted them to emerge. The bees have a plan (which almost certainly included further swarming) that you are now going to upset! Investigate the sealed cells carefully using a knife blade or scalpel and, if the queens are mature and ready to go, help several of them to walk out into the hive – the more the merrier (see **Figures 8 and 9**). The point is, that you do not know if there is already a virgin queen (or queens) loose in the hive so you are making sure by releasing the so-called `pulled virgins`. Having had your fun releasing virgin queens into the hive, you now have to do what it says in Step 5 and carefully destroy ALL the remaining queen cells (sealed or unsealed). Releasing all these queens into the hive at the same time sounds counter intuitive but seems to force the colony to select from the available talent and settle down to get their chosen queen mated. No matter how many queens are released (and we have released up to 18!) it does not seem to upset the colony and none has ever swarmed after following this procedure (yet!).

Figure 8: Emerged queen cell and releaseng virgin queen

STEP 8 – I think my hive has just produced a cast (secondary) swarm.

Investigation – This is very similar situation to **Step 7**. You need to find out if there are any un-emerged queen cells in the hive. You also need to look at any brood to find out how long it has been since the prime swarm departed.

Figure 9: Released virgin queen on comb

Remedial Action - If there are some un-emerged queen cells, do the same as in **Step 7** and release some virgin queens and then destroy all remaining queen cells. At this stage in the swarming process the beekeeper should be very suspicious of un-emerged queen cells because they are likely to be in this condition because the occupant is **DEAD**. If there are no un-emerged queen cells you have to decide whether there is a virgin queen in the hive? There probably is but, if you want to be sure, you can insert a `test` frame containing eggs and young larvae taken from another hive. If in a few days time there are no emergency queen cells on this frame, then there is a queen present and it simply a matter of waiting for her to start to lay. If queen cells are produced, then there is no queen and you can either allow these cells to produce a new queen or, if you have a source of a queen or sealed queen cells from another colony, you can introduce these to the hive and save time (until there is a new laying queen).

STEP 9 – My hive has no unsealed brood, a limited amount of sealed brood and no sealed queen cells.

Investigation – The age of the brood will give some idea when the original swarm occurred. Uncap a few cells and assess the age of the occupants. There should be a virgin queen in the hive so you need to ask yourself whether the colony is behaving queen-right; does it seem settled or are the bees running around, fanning their wings and making a `roaring` noise? Another question to ask, is there a laying arc? This is a semi-circle of cells on frames, usually in the middle of the hive, that have been kept free of nectar and are highly polished ready for a queen to lay in them. Neither of these `signs` are completely foolproof.

Remedial Action – There is not much you can do in this situation except check if the colony has a queen by inserting a `test` frame and see if queen cells are made (see **Step 8**). It is always advisable to do the test sooner rather than later because, if there is no queen, you have just been wasting time waiting for something to happen. If the message is in the affirmative (no queen cells are made) all you can do is wait for the queen to start to lay.

STEP 10 – My hive has no brood and no sealed queen cells, help!

Investigation – You now have very little information to tell you what has happened and when it happened. You may be able to see the remains of some queen cells but it will not be possible to tell how old these are. All possibility of swarming is now over and it is just a question as to whether or not this colony will get a new laying queen.

Remedial Action - Again you can use the `test` frame method to find out the queen status as described in **Steps 8 and 9** – sooner rather than later.

STEP 11 – My hive has no brood apart from that on a `test frame` it has received but NO queen cells have been produced.

Investigation – Failure to make queen cells on the `test frame` is because the colony thinks (or thought at the time you introduced the frame) that it had a queen – in other words there is (or was) a source of queen pheromone in the hive. The first question in this situation is how long is it since a queen cell could have emerged? There is probably no way you tell by looking at the frames so, if you have no information from previous inspections, you can not answer this question. If you do know the probable emergence date then it is reasonable to expect a laying queen in 3 weeks or, at the absolute limit, 4 weeks. Be aware that queens that are late coming into lay are subject to a higher failure rate (at some time in the future) than queens who start to lay on time. You can assess the behaviour of the colony as described in Step 9; is it behaving in a calm manner, does it have a laying arc? You can also look to see if you can see a queen but non-laying queens are difficult to find. You could give the colony another `test frame` but time is running out for them to be successful in producing a new queen.

Remedial Action – To be successful in re-queening by any method (using a a frame with eggs and young larvae, a sealed queen cell or a queen) you must find the source of queen pheromone and eliminate it. You need to search diligently for a queen and, if you find her, you must kill her – providing you are convinced she is never going to lay. If you have been successful in this task you will now be able to re-queen the colony. However, introducing a mature queen cell from another colony or a laying queen is likely to be a better option than a letting them raise their own queen at this late stage. Virgin queens are notoriously difficult to introduce but worth a try if that is all you have.

STEP 12 – My hive has got a drone-laying queen.

Investigation – A drone-laying queen can be identified by the presence of brood in worker cells that has domed capping – like the capping on drone cell but on a smaller worker cell. This means that the queen is laying unfertilised eggs when she should be laying fertilised ones. This can be due to a variety of reasons; she may not have mated properly, she may have run out of sperm or she may have some internal defect.

Initially, such a queen may lay both fertilised and unfertilised eggs, producing a mixture of normal and abnormal worker brood. Things can only get worse (not better) so now is the time to take action.

Remedial Action – This is exactly the same as for **Step 11** – you must find the queen and eliminate her before you can re-queen. Occasionally, worker brood with domed cappings is not produced by a drone-laying queen but by laying workers. This is not easy to diagnose. The signature of laying workers is that the brood is patchy with little pattern, eggs may be laid on the side of the cells and there may be more than on egg/cell. It is said that a colony with laying workers will not accept a mature queen cell or even a laying queen but usually there is no problem. If the colony is still worth saving (or rather the bees in it are worth saving) the safe option is to unite it with a queen-right colony.

Postscript

If you have diligently followed the steps in the above key, at each stage making a careful study of the evidence available in the hive, you should now be in a position to manage that colony with the best chance of a favourable outcome – for both the bees and the beekeeper. You will not always be successful, either because you have failed to arrive at the correct diagnosis (it is not always easy) or because of matters outside of your control, such as queen mating, but you will have given it your best shot. Honey bee colonies are individuals and this is part of the challenge and fascination of beekeeping.

Appendix 1

Not Sure Whether the Queen Cells are for Swarming or Supersedure

If the cells are sealed or near sealing and a swarm could be imminent then the safest option is to split the colony. The queen should be left where she is and the queen cells along with several frames of brood (3-4) and sufficient bees to look after them should be put in a nuc box or on a split board at the top of the hive. The queen cells in the split will mature and in due course a virgin queen will emerge and because all the flying bees have returned to their original site (where the queen is) no swarming will occur. If left in this condition the new queen will mate and start to lay. It is subsequent events in the main hive that need to be closely followed. If a new batch of queen cells is immediately started (and there may be more of them this time) then it is almost certain that the original intention was swarming and the colony will need to be artificially swarmed. If no more queen cells are produced then it was probably an attempt at supersedure. In this case the performance of the 'suspect' queen needs to be monitored; is she producing a satisfactory amount of brood? Further supersedure cells may or may not be produced in the short term. If the queen is still performing well then things can be left until the new queen is up and running and then the decision can be made whether (or not) she should be used to replace her mother as the colony originally intended.

Appendix 2: Failure of the Prime Swarm

The practice of wing-clipping the queen has already been discussed above. Clipping causes failure of the prime swarm but the beekeeper should be aware that this can occur naturally from time to time. It happens when the old queen, for reasons that are not usually obvious, is unable or unwilling to fly, she falls to the ground and the prime swarm is aborted in favour of the first virgin to emerge. Often there is little evidence as to what has happened, except that you appear to have a colony that has swarmed (it has no queen) but does not seem to have lost any bees – which is good! There is another possible outcome to this situation about which you need to be aware. The non-flying queen will sometimes crawl under a hive stand or similar refuge and the swarm will try to establish themselves in an unsuitable (for them and the beekeeper) place. The swarm can be recovered from where is has settled and re-housed in a hive. But, if it has been there for some time before it was noticed, it will have re-located itself and it will be necessary to remove it to a new location – theoretically at least 3 miles away but a shorter distance will usually suffice. The remedial action for the parent colony (the one from which the swarm has issued) is the same in all cases; the queen cells must be thinned in order to prevent a cast swarm (see **Step 5** for details) or the date for queen cells maturity can be estimated in order to do some 'virgin pulling' (see **Step 7** for details).

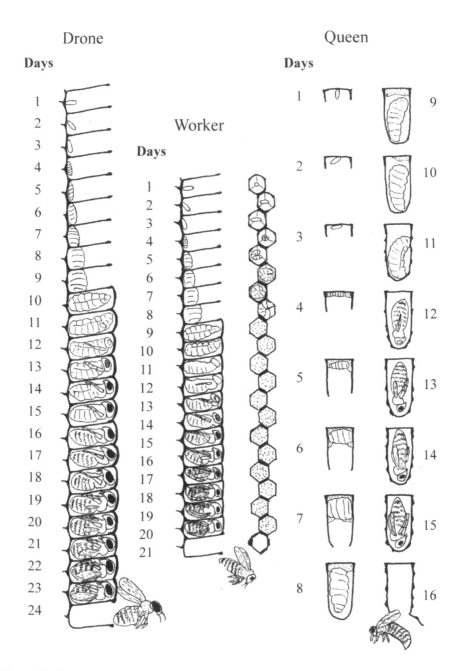

Figure 10: See page 10

Lightning Source UK Ltd.
Milton Keynes UK
UKHW050742050422
401073UK00003B/19